North America

North America presents an overview of the geography of this continent and the countries that make up North America. The teaching and learning in this unit are based on the five themes of geography developed by the Association of American Geographers together with the National Council for Geographic Education.

The five themes of geography are described on pages 2 and 3. The themes are also identified on all student worksheets throughout the unit.

North America is divided into seven sections.

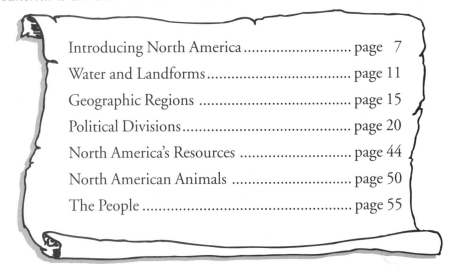

Each section includes:
    teacher resource pages explaining the activities in the section
    information pages for teachers and students
    reproducible resources
        maps
        note takers
        activity pages

Pages 4–6 provide suggestions on how to use this unit, including instructions for creating a geography center.

This book has been correlated to over 145 state standards.

www. teaching-standards.com

Visit this site to view a correlation of this book's activities to your state's standards. This is a free service.

EMC 763

Evan-Moor
EDUCATIONAL PUBLISHERS

Author: Jo Ellen Moore
Editor: Jill Norris
Copy Editor: Cathy Harber
Desktop: Keli Winters
Illustrators: Cindy Davis
                Keli Winters
Cover: Cheryl Puckett
Photography: David Bridge
                and Digital Stock

Congratulations on your purchase of some of the finest teaching materials in the world.

For information about other Evan-Moor products, call 1-800-777-4362 or FAX 1-800-777-4332. Visit our Web site www.evan-moor.com. Check the Product Updates link for supplements, additions, and corrections for this book.

The Five Themes of Geography

Location

Position on the Earth's Surface

Location can be described in two ways. **Relative location** refers to the location of a place in relation to another place. **Absolute location** (exact location) is usually expressed in degrees of longitude and latitude.

> We can say the U.S.A. is located in the northern hemisphere between the Pacific and the Atlantic Oceans. Mexico and Canada form its other two borders.

> Washington, D.C., is located at 39°N latitude, 77°W longitude.

Place

Physical and Human Characteristics

Place is expressed in the characteristics that distinguish a location. It can be described in **physical characteristics** such as water and landforms, climate, etc., or in **human characteristics** such as languages spoken, religion, government, etc.

> Much of Alaska and northern Canada are tundra. The tundra is a vast, treeless plain where the ground beneath the surface stays frozen even in the summer.

> English is the official language of most of the North American countries.

Relationships within Places

Humans and the Environment

This theme includes studies of how people depend on the environment, how people adapt to and change the environment, and the impact of technology on the environment. Cities, roads, planted fields, and terraced hillsides are all examples of man's mark on a place. A place's mark on man is reflected in the kind of homes built, the clothing worn, the work done, and the foods eaten.

> Many of the plains of central North America are covered in fields of wheat. Before farmers planted the wheat, the plains were covered with wild grasses, and herds of buffalo and antelope grazed there.

Movement

Human Interactions on the Earth

Movement describes and analyzes the changing patterns caused by human interactions on the Earth's surface. Everything moves. People migrate, goods are transported, and ideas are exchanged. Modern technology connects people worldwide through advanced forms of communication.

Use of the Internet to communicate with the rest of the world is becoming popular throughout North America.

Immigrants from all over the world continue to come to North America in search of a better life for their families.

Regions

How They Form and Change

Regions are a way to describe and compare places. A region is defined by its common characteristics and/or features. It might be a geographic region, an economic region, or a cultural region.

Geographic region: The Chihuahuan Desert covers a large part of northern Mexico.
Economic region: Tourists are the major source of income for the Hawaiian Islands.
Cultural region: Although English and French are both official languages of Canada, French is spoken for the most part in Quebec.

Using This Geography Unit

Good Teaching with *North America*

Use your everyday good teaching practices as you present material in this unit.

* Provide necessary background and assess student readiness:
    review necessary skills such as using latitude, longitude, and map scales
    model new activities
    preview available resources
* Define the task on the worksheet or the research project:
    explain expectations for the completed task
    discuss evaluation of the project
* Guide student research:
    provide adequate time for work
    provide appropriate resources
* Share completed projects and new learnings:
    correct misconceptions and misinformation
    discuss and analyze information

Doing Student Worksheets

Before assigning student worksheets, decide how to manage the resources that you have available. Consider the following scenarios for doing a page that requires almanac or atlas research:

* You have one classroom almanac or atlas.
    Make an overhead transparency of the page needed and work as a class to complete the activity, or reproduce the appropriate almanac page for individual students. (Be sure to check copyright notations before reproducing pages.)
* You have several almanacs or atlases.
    Students work in small groups with one resource per group, or rotate students through a center to complete the work.
* You have a class set of almanacs or atlases.
    Students work independently with their own resources.

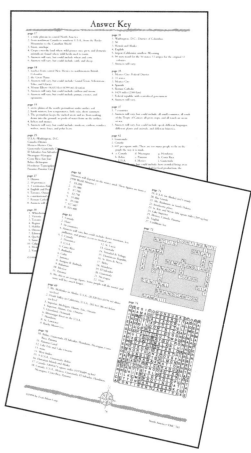

Checking Student Work

A partial answer key is provided on pages 77 and 78. Consider the following options for checking the pages:

* Collect the pages and check them yourself. Then have students make corrections.
* Have students work in pairs to check and correct information.
* Discuss and correct the pages as a class.

Creating a Geography Center

Students will use the center to locate information and to display their work.

Preparation

1. Post the unit map of North America on an accessible bulletin board.
2. Add a chart for listing facts about North America as they are learned.
3. Allow space for students to display newspaper and magazine articles on the continent, as well as samples of their completed projects.
4. Provide the following research resources:
   * world map
   * globe
   * atlas (one or more)
   * current almanac
   * computer programs and other electronic resources
   * fiction and nonfiction books (See bibliography on pages 79 and 80.)
5. Provide copies of the search cards (pages 69–71), crossword puzzle (pages 72 and 73), and word search (page 74). Place these items in the center, along with paper and pencils.

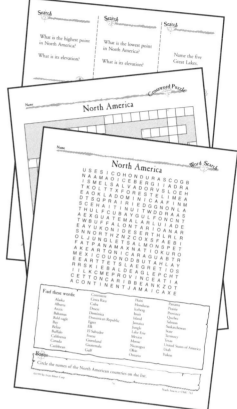

Additional Resources

At appropriate times during the unit, you will want to provide student access to these additional research resources:
   * Filmstrips, videos, and laser discs
   * Bookmarked sites on the World Wide Web (For suggestions, go to http://www.evan-moor.com and click on the Product Updates link on the home page.)

Making a Portfolio on North America

Provide a folder in which students save the work completed in this unit.
Reproduce the following portfolio pages for each student:

* A Summary of Facts about North America, page 66
  Students will use this fact sheet to summarize basic
  information they have learned about North America.
  They will add to the sheet as they move through
  the unit.

* What's Inside This Portfolio?, page 67
  Students will record pages and projects that they
  add to the portfolio, the date of each addition,
  and why it was included.

* My Bibliography, page 68
  Students will record the books and other materials
  they use throughout their study of North America.

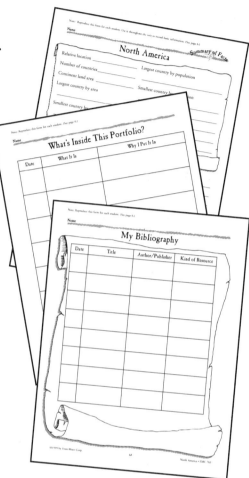

At the end of the unit have students create a cover
illustration showing some aspect of North America.

Encourage students to refer to their portfolios often.
Meet with them individually to discuss their learning.
Use the completed portfolio as an assessment tool.

Using the Unit Map

Remove the full-color unit map from the center of this book and use it to help students
do the following:

* locate and learn the names of landforms, water forms, and physical regions of
  North America
* practice finding relative locations using cardinal directions shown on the compass rose
* calculate distances between places using the scale

Introducing North America

Tour the Geography Center

Introduce the Geography Center to your class. Show the research materials and explain their uses. Ask students to locate the sections of atlases and almanacs containing material about North America.

Thinking about North America

Prepare a KWL chart in advance. Reproduce page 8 for each student. Give students a period of time (5–10 minutes) to list facts they already know about North America and questions about the continent they would like answered.

Half Dome at Yosemite National Park

Know	Want to Know	Learned

Transfer their responses to the KWL chart. Post the chart in a place where you can add to it throughout your study of the continent.

Where Is North America?

Reproduce pages 9 and 10 for each student.

"Locating North America" helps students locate North America using relative location. Use the introductory paragraph to review the definition of relative location, and then have students complete the page.

"Name the Hemisphere" reviews the Earth's division into hemispheres. Students are asked to name the hemispheres in which North America is located. Using a globe to demonstrate the divisions, read the introduction together. Then have students complete the page.

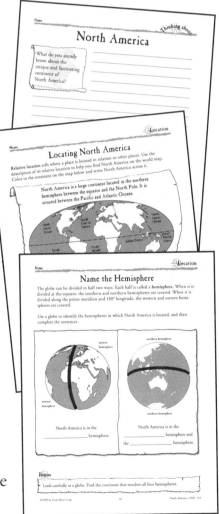

North America

What do you already know about the unique and fascinating continent of North America?

_____

_____

_____

_____

_____

_____

_____

_____

If you could talk to someone from North America, what would you ask?

_____

_____

_____

_____

_____

_____

_____

_____

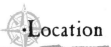

Locating North America

Relative location tells where a place is located in relation to other places. Use the description of its relative location to help you find North America on the world map. Color in the continent on the map below and write North America across it.

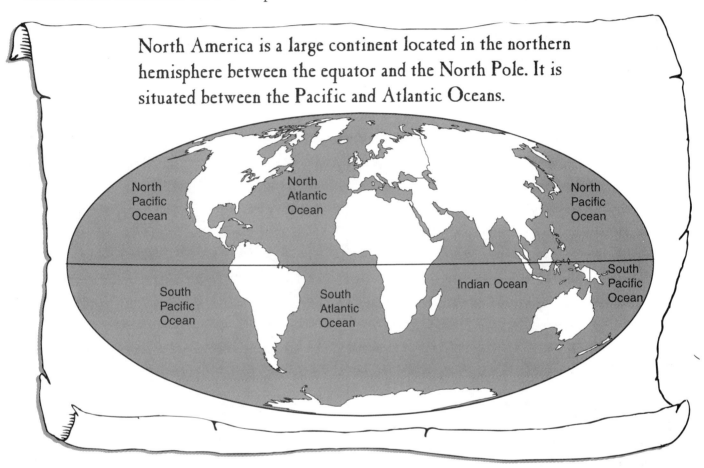

North America is a large continent located in the northern hemisphere between the equator and the North Pole. It is situated between the Pacific and Atlantic Oceans.

North Pacific Ocean

North Atlantic Ocean

North Pacific Ocean

South Pacific Ocean

South Atlantic Ocean

Indian Ocean

South Pacific Ocean

Look at a map of North America. Find these places and write their relative locations:

1. Cuba _____

2. Canada _____

3. El Salvador _____

Bonus

Write the relative location of the school you attend.

Name the Hemisphere

The globe can be divided in half two ways. Each half is called a **hemisphere**. When it is divided at the equator, the southern and northern hemispheres are created. When it is divided along the prime meridian and 180° longitude, the western and eastern hemispheres are created.

Use a globe to identify the hemispheres in which North America is located, and then complete the sentences.

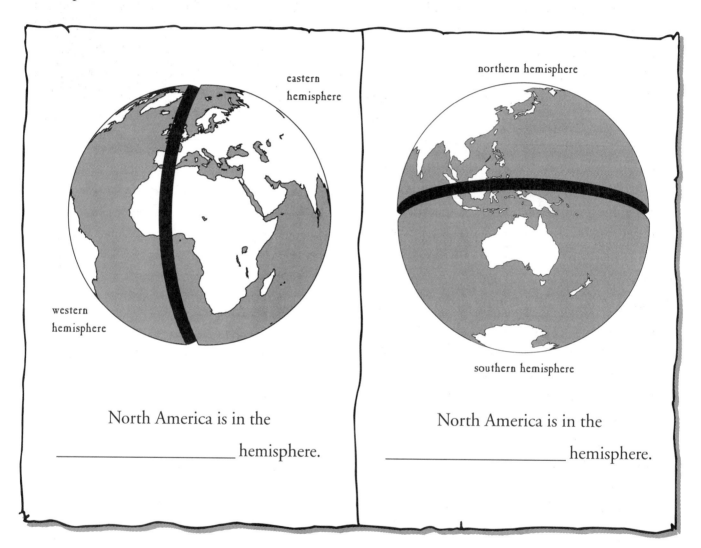

North America is in the

_____ hemisphere.

North America is in the

_____ hemisphere.

Bonus

Look carefully at a globe. Find the continent that touches all four hemispheres.

Water and Landforms

Collecting information by reading physical maps involves many skills. Pages 12–14 provide students with the opportunity to refine these skills as they learn about the water and landforms on the continent of North America.

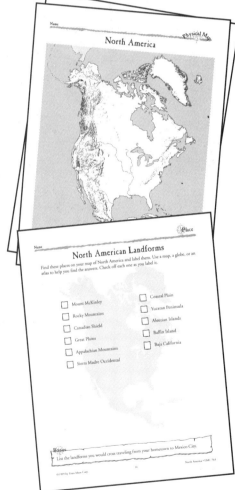

Bryce Canyon, Utah

Water Forms

Reproduce pages 12 and 13 for each student. Use the unit map to practice locating oceans, seas, lakes, and rivers on a map.

* Review how rivers and lakes are shown on a map.
* Discuss pitfalls students may face in finding the correct names (names written along the rivers, small type, several names close together).
* Have students locate at least one example of each type of water form on the unit map.
* Then have students locate and label the listed water forms on their individual physical maps.

Landforms

Reproduce page 14 for each student. Have students use the same map used to complete page 13, or reproduce new copies of page 12 for this activity.

* Review the ways mountains, deserts, and other landforms are shown on a map (symbols, color variations, labels).
* Have students practice locating some of the mountains, deserts, and other landforms on the unit map of North America.
* Then have students locate and label the listed landforms on their individual physical maps.

North America

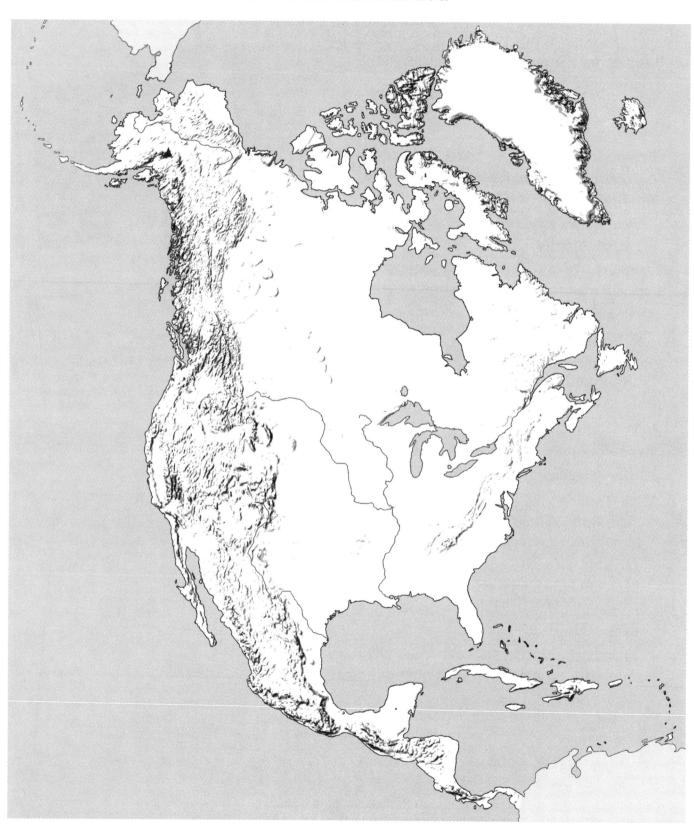

Water Forms of North America

Find these places on your map of North America and label them. Use a map, a globe, or an atlas to help you find the answers. Check off each one as you label it.

☐ Pacific Ocean   ☐ Mississippi River

☐ Atlantic Ocean   ☐ St. Lawrence River

☐ Arctic Ocean   ☐ Gulf of California

☐ Great Lakes   ☐ Hudson Strait

☐ Gulf of Mexico   ☐ Missouri River

☐ Caribbean Sea   ☐ Rio Grande

☐ Bering Sea   ☐ Yukon River

☐ Hudson Bay   ☐ Baffin Bay

Trace the rivers dark blue.

Color the lakes dark blue.

Color the seas, oceans, and gulfs light blue.

Bonus

Imagine you are traveling in a small ship. Explain the route you would follow to get from Baffin Island to the Panama Canal.

North American Landforms

Find these places on your map of North America and label them. Use a map, a globe, or an atlas to help you find the answers. Check off each one as you label it.

☐ Mount McKinley   ☐ Coastal Plain

☐ Rocky Mountains   ☐ Yucatan Peninsula

☐ Canadian Shield   ☐ Aleutian Islands

☐ Great Plains   ☐ Baffin Island

☐ Appalachian Mountains   ☐ Baja California

☐ Sierra Madre Occidental

Bonus

List the landforms you would cross traveling from your hometown to Mexico City.

Geographic Regions

The continent of North America is large, with many different geographic regions. Each region has distinct physical characteristics and climatic conditions. The material on pages 16–19 explores some of these regions.

Regions of North America

Reproduce pages 16–19 for each student. As a class, discuss the material about physical regions, referring to the unit map to locate each region. Share additional information from books and videos in your geography center. Students use class resources to find the answers to the questions on the activity pages. (You may have all students do each page, or divide the class into three groups, each doing research on one region.)

- Great Plains, page 17
- Rocky Mountains, page 18
- Arctic Tundra, page 19

Comparing Regions

Prepare a large chart on butcher paper (see below). Have students work together to fill in each box using the information gathered as they study the various regions. Do additional research using class resources to fill in any missing information. Then have each student select one of the regions on the chart, synthesize the information that has been gathered, and write a report about the region.

	Great Plains	Rocky Mountains	Arctic Tundra
Location			
Climate			
Plants living in the region			
Animals living in the region			
People living in the region			
Ways people have changed the region			

Extend the lesson by challenging students to explore the other regions found in North America. Explain that deserts and forests of many kinds are found on the continent and its island countries. Allow time for students to share their information with the rest of the class.

Regions of North America

The continent of North America has many geographic regions. There are high mountains and rolling hills, vast deserts, wide plains, and many lakes and rivers.

Some regions, such as the Canadian Shield, cover large areas. The Canadian Shield is a plateau covering the eastern half of Canada, most of Greenland, and sections of northern Minnesota, Wisconsin, Michigan, and New York.

Another large region is a coastal plain reaching from the eastern United States down into Mexico. To the west of this coastal plain is a narrow range of mountains and hills.

The central part of the continent, from southern Canada to southwest Texas, is a lowland that contains some hilly terrain. The western part of this flatland consists of the Great Plains, which slope up to the foot of the Rocky Mountains.

The Great Plains

The Rocky Mountains

The Rocky Mountains of the United States and Canada are related to the Sierra Madre Oriental range in Mexico. Along the Pacific coast are several mountain ranges that reach from the Alaska Range to the Sierra Madre Occidental in Mexico.

Each of these regions has plant and animal life adapted to the climate and terrain. Parts of each region have been given specific names.

Bonus

Learn more about other regions of North America such as the Chihuahuan Desert, the Great Basin, the Florida Everglades, or the tropical rainforests of Panama and Nicaragua.

Great Plains

One major region of North America is the Great Plains, or Central Plains. This is a wide plateau in central North America. The plains reach from northwestern Canada into the United States as far south as Texas. They are bordered on the west by the Rocky Mountains and on the east by the Canadian Shield.

Originally, the Great Plains were vast grasslands with herds of grazing animals. Today much of the land is used for commercial crops and livestock farms. Coal deposits and underground fields of petroleum are important energy resources.

Answer these questions:

1. What are the Great Plains? _____

_____

2. Where are they located? _____

_____

3. What types of animals used to roam the grasslands of the Great Plains?

_____

_____

4. How has human activity changed the Great Plains? _____

_____

5. What types of crops are raised there today? _____

_____

6. What types of livestock are raised there? _____

_____

Bonus
What happened to the native people of the Great Plains when European settlers moved onto the land?

Rocky Mountains

A long mountain range reaches from New Mexico to British Columbia along the western side of the continent. The Rocky Mountains form part of what is called the Continental Divide. These mountain ranges separate the rivers draining into the Atlantic or Arctic Oceans from those flowing into the Pacific Ocean. Many of the mountain peaks in the Rocky Mountains are more than 14,000 feet (4267 m) high. The highest mountain in the range is Mount Elbert, located in central Colorado.

The lower levels of the mountains are covered in grasslands. As you move up the mountains you find forests. Above the timberline are grasses and scattered shrubs, and at the summit there is very little vegetation. Some of the highest peaks are covered in snow and ice all year round.

Answer these questions:

1. Describe the range of the Rocky Mountains. _____

_____

2. What North American region borders the Rocky Mountains on its eastern side?

_____

3. Name two of the national parks located in the Rocky Mountains.

_____

4. Name the highest mountain in the Rocky Mountain range and give its elevation.

_____

5. What kinds of animals are found in the northern part of the Rocky Mountains?

_____

6. What kinds of animals are found in the southern part of the Rocky Mountains?

_____

Bonus

What is the highest mountain in North America? Is it part of the Rocky Mountains?

Arctic Tundra

The Arctic plains, called the tundra, are found in northern Canada and in the state of Alaska. The tundra is an example of a cold desert. It receives very little precipitation. The winters are long and harsh with very low temperatures, and the summers are short.

The Arctic tundra has a layer of permanently frozen subsoil about 3 feet (about 1 m) under the surface. Even in the summer, this layer does not thaw. When the surface snow and ice melt, the water cannot drain through the layer of permafrost. It stays on the surface of the tundra in bogs and ponds.

The region is cold and windy even in the summer. In spite of this, there are plants that grow on the tundra. Mosses and lichens are the most common plants, but there are also many varieties of flowering plants that bloom during the brief growing season. Flocks of birds, swarms of insects, and larger animals all take advantage of the warmer days and greater availability of food plants.

Answer these questions:

1. What is the tundra? _____

_____

2. Describe the usual weather conditions. _____

_____

3. Why do bogs and pools form on the surface of the tundra during the summer?

_____

_____

4. Name the most common plants growing on the tundra. _____

_____

5. Name some of the mammals that live on the tundra. _____

_____

Bonus
Most birds migrate to warmer areas during the winter. Name one bird that stays on the tundra all year. Name three animals that change color from white for the winter to brown for the short summer.

Political Divisions

A political map shows boundaries between countries or between states, provinces, and territories. In this section students will use political maps to learn the countries of North America and their capital cities, to calculate distance and direction, and to locate places using longitude and latitude.

The Countries of North America

North America

Reproduce pages 23–25 for each student. Have students use map resources to list the countries and then label them on the political map. Then have students do the activity on page 25. (Students will label island countries on page 40.)

Note: Many Central American countries are small. Have students explore ways they might label these countries (draw a line from the country to open space and write the name there, number or color-code the small countries and make a key at the bottom of the map, write the names of island countries in the water next to them, etc.).

Using a Compass Rose

Reproduce page 26 for each student. Use the compass rose on the unit map to review how to determine location using cardinal directions. Then have students use their political maps to complete the activity independently.

Canada

Provinces and Territories

Reproduce pages 27 and 28 for each student. As a class, read and discuss the information, and then have students use class resources to answer the questions. Students then use map resources to name the provinces and territories and their capital cities and label them on the map.

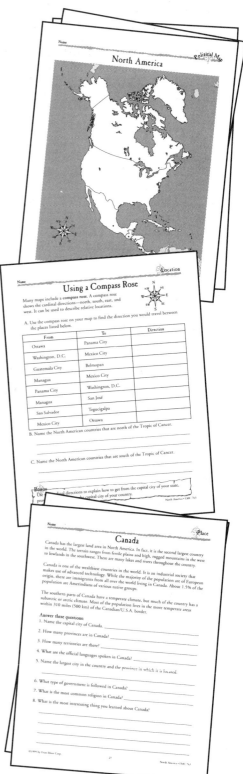

Nunavut

Reproduce page 29 for each student. Make an overhead transparency of the map section of the page. Explain that while the political boundaries of places in North America have been stable for many years, changes occasionally do occur. As a class, read and discuss the change that just occurred (April 1999) in the Northwest Territories of Canada. Have students use class resources to find out more about the formation of Nunavut. (Check the World Wide Web for sites students can use to find out more about this area.)

Longitude and Latitude

Reproduce page 30 for each student. Use a unit map to review how to use lines of longitude and latitude to determine exact locations. Then have students complete the activity independently, using their political maps of Canada (page 28).

The United States of America

States and Capital Cities

Reproduce pages 31 and 32 for each student. As a class, read and discuss the information, and then have students use class resources to answer the questions. Students then use map resources to name the states and their capital cities and label them on the map.

How Far Is It?

Reproduce page 33 for each student. Use the unit map to review how to use a map scale to figure distances. Then have students use a ruler and the map scale to determine the distance between various places in the United States of America (page 32).

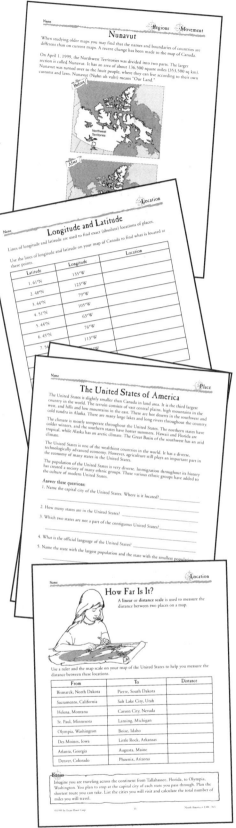

Comparison Chart

Reproduce page 34 for each student. Have students research two diverse states, provinces, or territories in Canada and the United States. (This is a good opportunity to use the school and public libraries, as well as electronic resources.) Assign one of the following or have students select two places on their own:

* Alaska and Hawaii
* New York and New Mexico
* Quebec and Yukon
* Manitoba and Newfoundland
* Wisconsin and California
* Ontario and Nunavut

Other Countries in North America

Reproduce pages 35–40 for each student. As a class, read and discuss the information on each page, and then have students use class resources to answer the questions. Students then follow directions to complete the maps of Mexico, Central America, and the island countries.

Mexico, pages 35 and 36
Central America, pages 37 and 38
The Island Countries, pages 39 and 40

How Many People Live Here?

There is a big difference in population density among the countries in North America. Make an overhead transparency of page 41. Reproduce pages 41 and 42 for each student. Show the graph as students discuss the population density of the various countries shown. Ask students to think of reasons why a large country such as Canada has so few people per square mile, while a small country such as Costa Rica has so many. Challenge students to think of ways population density affects the lives of the people living there.

Country Fact Sheet

Reproduce page 43 for each student. Explain that students will help create a file of fact sheets for the countries of North America. Have each student select a different country to research. Allow time for students to share what they discover about their countries. Keep the completed sheets in a binder in the geography center.

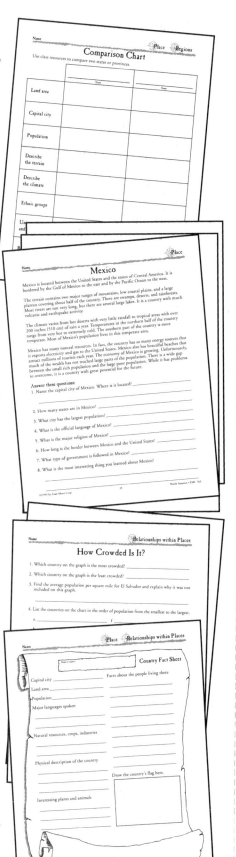

North America

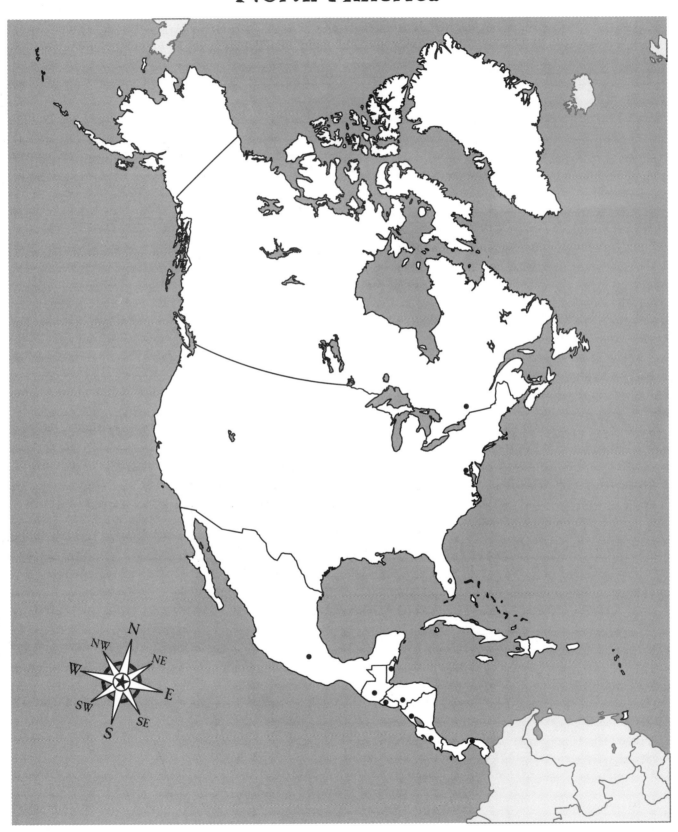

The Countries of North America

List the name of each country that is part of the North American continent. (Don't forget the island countries.) Locate and label on your political map each of the 10 countries that are part of the North American mainland.

_____   _____

_____   _____

_____   _____

_____   _____

_____   _____

_____   _____

_____   _____

_____   _____

_____   _____

_____

Capital Cities

Complete this chart by listing the capital cities of these North American countries.

Country	Capital City
United States of America	
Canada	
Mexico	
Guatemala	
El Salvador	
Nicaragua	
Costa Rica	
Belize	
Honduras	
Panama	

Now write each capital city in the correct place on your map.

Bonus
Write the capital city of your country and the capital city of the state, province, or territory in which you live.

Using a Compass Rose

Many maps include a **compass rose**. A compass rose shows the cardinal directions—north, south, east, and west. It can be used to describe relative locations.

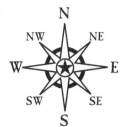

A. Use the compass rose on your map to find the direction you would travel between the places listed below.

From	To	Direction
Ottawa	Panama City	
Washington, D.C.	Mexico City	
Guatemala City	Belmopan	
Managua	Mexico City	
Panama City	Washington, D.C.	
Managua	San José	
San Salvador	Tegucigalpa	
Mexico City	Ottawa	

B. Name the North American countries that are north of the Tropic of Cancer.

_____

_____

C. Name the North American countries that are south of the Tropic of Cancer.

_____

_____

Bonus

Use the cardinal directions to explain how to get from the capital city of your state, province, or territory to the capital city of your country.

Canada

Canada has the largest land area in North America. In fact, it is the second largest country in the world. The terrain ranges from fertile plains and high, rugged mountains in the west to lowlands in the southwest. There are many lakes and rivers throughout the country.

Canada is one of the wealthiest countries in the world. It is an industrial society that makes use of advanced technology. While the majority of the population are of European origin, there are immigrants from all over the world living in Canada. About 1.5% of the population are Amerindians of various native groups.

The southern parts of Canada have a temperate climate, but much of the country has a subarctic or arctic climate. Most of the population lives in the more temperate areas within 310 miles (500 km) of the Canadian/U.S.A. border.

Answer these questions:

1. Name the capital city of Canada. _____

2. How many provinces are in Canada? _____

3. How many territories are there? _____

4. What are the official languages spoken in Canada? _____

5. Name the largest city in the country and the province in which it is located.

_____

6. What type of government is followed in Canada? _____

7. What is the most common religion in Canada? _____

8. What is the most interesting thing you learned about Canada?

_____

_____

_____

_____

Name

Canada

Label each province and territory. Then write the names of the capital cities.

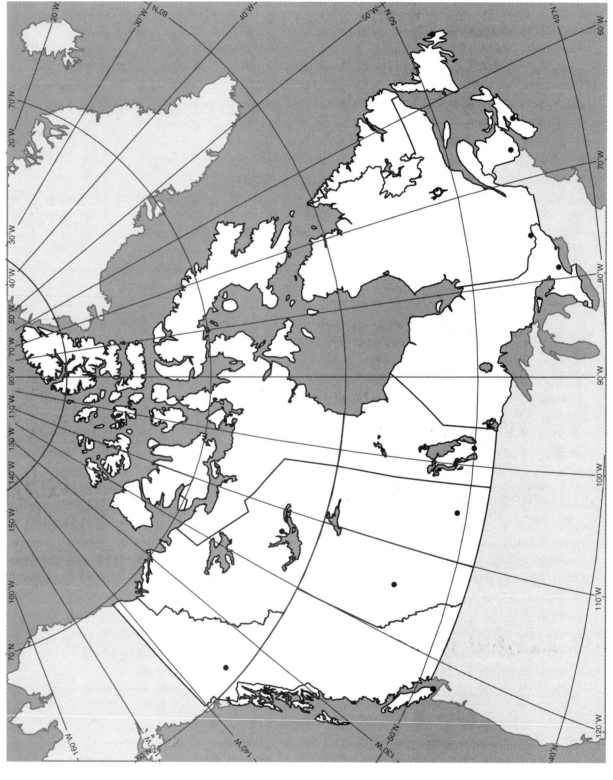

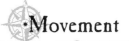 
Nunavut

When studying older maps you may find that the names and boundaries of countries are different than on current maps. A recent change has been made to the map of Canada.

On April 1, 1999, the Northwest Territories was divided into two parts. The larger section is called Nunavut. It has an area of about 136,500 square miles (353,500 sq km). Nunavut was turned over to the Inuit people, where they can live according to their own customs and laws. Nunavut (Nuhn uh vuht) means "Our Land."

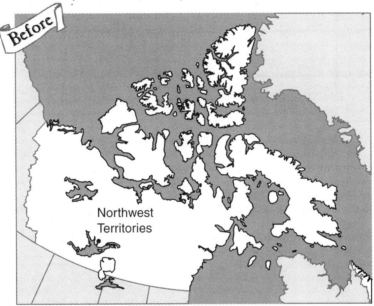

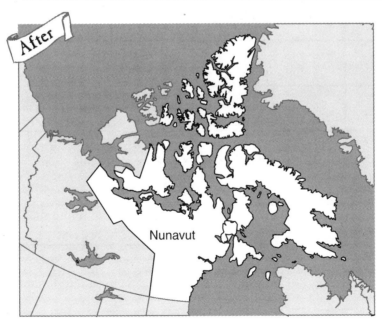

Use class resources to find out more about the new territory and its people. Share what you learn with your classmates.

Longitude and Latitude

Lines of longitude and latitude are used to find exact (absolute) locations of places.

Use the lines of longitude and latitude on your map of Canada to find what is located at these points.

Latitude	Longitude	Location
1. 61°N	135°W	
2. 48°N	123°W	
3. 44°N	79°W	
4. 51°N	105°W	
5. 44°N	63°W	
6. 45°N	76°W	
7. 54°N	113°W	
8. 47°N	71°W	
9. 48°N	53°W	
10. 51°N	114°W	

Now write the latitude and longitude of these cities:

1. Montréal _____

2. Vancouver _____

Bonus

Write the longitude and latitude of the capital city of the country in which you live.

The United States of America

The United States is slightly smaller than Canada in land area. It is the third largest country in the world. The terrain consists of vast central plains, high mountains in the west, and hills and low mountains in the east. There are hot deserts in the southwest and cold tundra in Alaska. There are many large lakes and long rivers throughout the country.

The climate is mostly temperate throughout the United States. The northern states have colder winters, and the southern states have hotter summers. Hawaii and Florida are tropical, while Alaska has an arctic climate. The Great Basin of the southwest has an arid climate.

The United States is one of the wealthiest countries in the world. It has a diverse, technologically advanced economy. However, agriculture still plays an important part in the economy of many states in the United States.

The population of the United States is very diverse. Immigration throughout its history has created a society of many ethnic groups. These various ethnic groups have added to the culture of modern United States.

Answer these questions:

1. Name the capital city of the United States. Where is it located? _____

_____

2. How many states are in the United States? _____

3. Which two states are not a part of the contiguous United States? _____

_____

4. What is the official language of the United States? _____

5. Name the state with the largest population and the state with the smallest population.

_____

6. What do the stars and stripes on the flag represent? _____

_____

7. What is the most interesting thing you learned about the United States? _____

_____

Name

The United States of America

Label each state and its capital city.

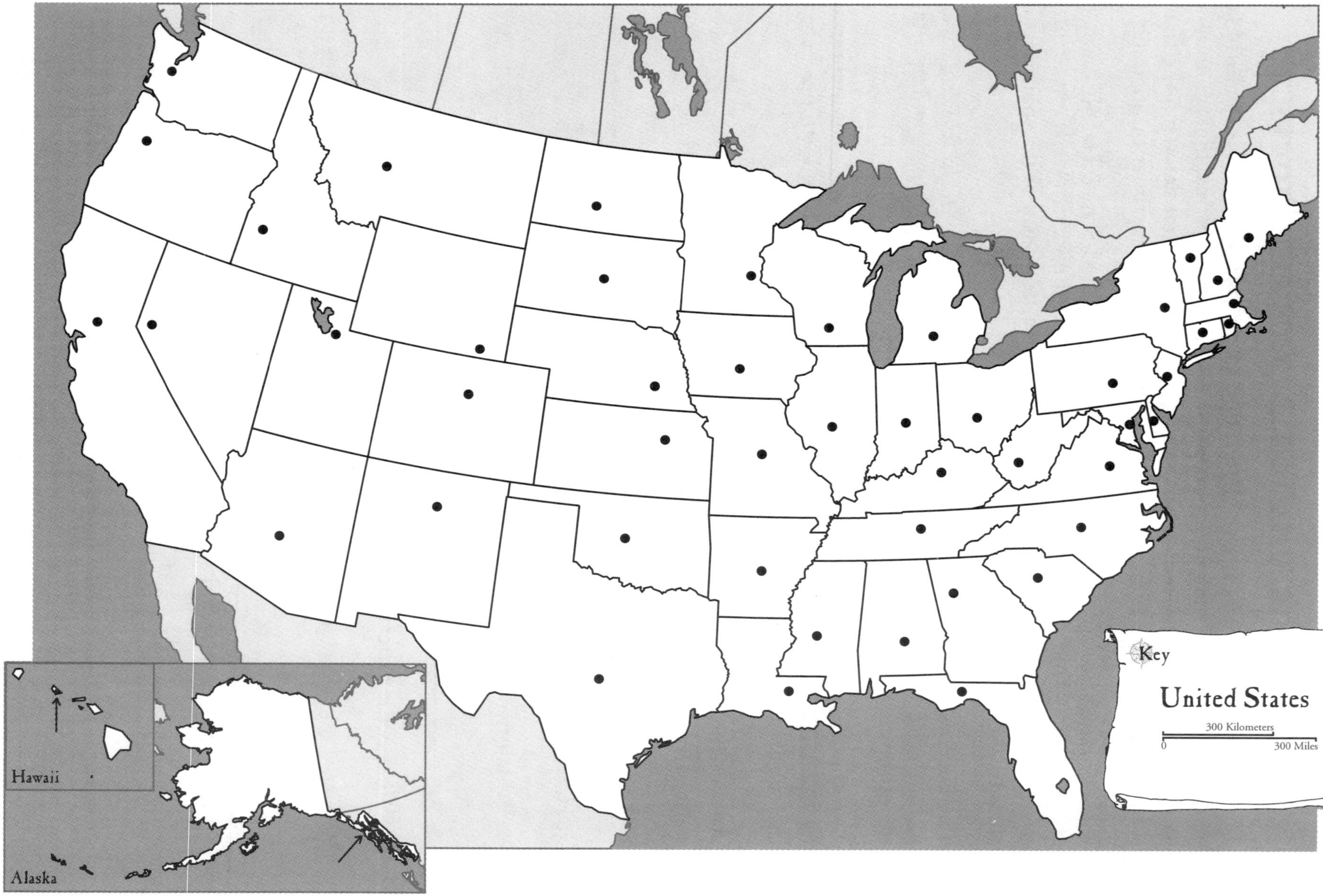

United States

Key

300 Kilometers

300 Miles

0

North America

Hawaii

Alaska

How Far Is It?

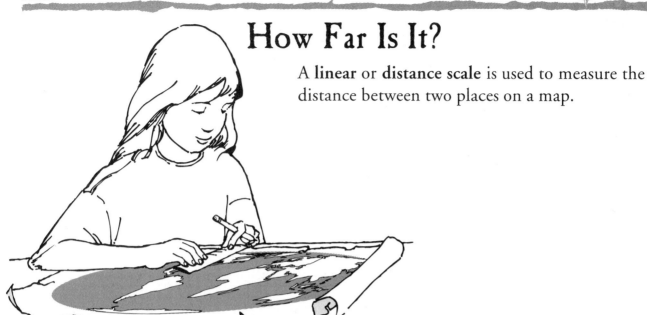

A **linear** or **distance scale** is used to measure the distance between two places on a map.

Use a ruler and the map scale on your map of the United States to help you measure the distance between these locations.

From	To	Distance
Bismarck, North Dakota	Pierre, South Dakota	
Sacramento, California	Salt Lake City, Utah	
Helena, Montana	Carson City, Nevada	
St. Paul, Minnesota	Lansing, Michigan	
Olympia, Washington	Boise, Idaho	
Des Moines, Iowa	Little Rock, Arkansas	
Atlanta, Georgia	Augusta, Maine	
Denver, Colorado	Phoenix, Arizona	

Bonus

Imagine you are traveling across the continent from Tallahassee, Florida, to Olympia, Washington. You plan to stop at the capital city of each state you pass through. Plan the shortest route you can take. List the cities you will visit and calculate the total number of miles you will travel.

Name

Comparison Chart

Use class resources to compare two states or provinces.

	Name	Name
Land area		
Capital city		
Population		
Describe the terrain		
Describe the climate		
Ethnic groups		
Unique plants and animals		
Resources		

•Place

Mexico

Mexico is located between the United States and the states of Central America. It is bordered by the Gulf of Mexico to the east and by the Pacific Ocean to the west.

The terrain contains two major ranges of mountains, low coastal plains, and a large plateau covering about half of the country. There are swamps, deserts, and rainforests. Most rivers are not very long, but there are several large lakes. It is a country with much volcanic and earthquake activity.

The climate varies from hot deserts with very little rainfall to tropical areas with over 200 inches (510 cm) of rain a year. Temperatures in the northern half of the country range from very hot to extremely cold. The southern part of the country is more temperate. Most of Mexico's population lives in this temperate area.

Mexico has many natural resources. In fact, the country has so many energy sources that it exports electricity and gas to the United States. Mexico also has beautiful beaches that attract millions of tourists each year. The economy of Mexico is growing. Unfortunately, much of the wealth has not reached large parts of the population. There is a wide gap between the small rich population and the large poor population. While it has problems to overcome, it is a country with great potential for the future.

Answer these questions:

1. Name the capital city of Mexico. Where is it located? _____

_____

2. How many states are in Mexico? _____

3. What city has the largest population? _____

4. What is the official language of Mexico? _____

5. What is the major religion of Mexico? _____

6. How long is the border between Mexico and the United States? _____

7. What type of government is followed in Mexico? _____

8. What is the most interesting thing you learned about Mexico?

_____

_____

Mexico

Label the states.

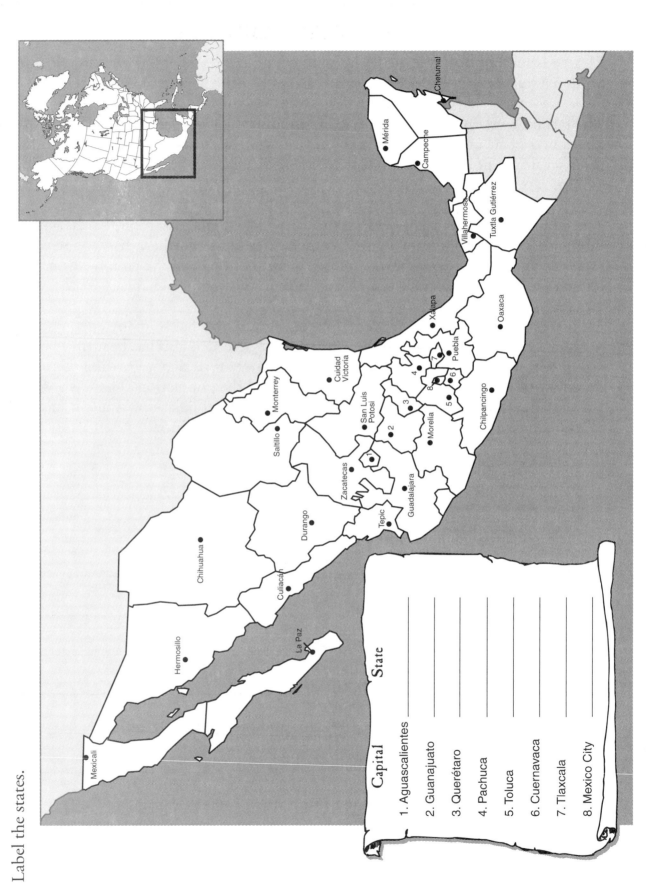

Mexicali

Hermosillo

La Paz

Chihuahua

Culiacán

Durango

Tepic

Zacatecas

Guadalajara

Saltillo

Monterrey

San Luis Potosí

Cuidad Victoria

Morelia

1

2

3

4

5

6

7

8

Chilpancingo

Puebla

Oaxaca

Xalapa

Villahermosa

Tuxtla Gutiérrez

Mérida

Campeche

Chetumal

Capital   State

1. Aguascalientes _____

2. Guanajuato _____

3. Querétaro _____

4. Pachuca _____

5. Toluca _____

6. Cuernavaca _____

7. Tlaxcala _____

8. Mexico City _____

·Place

Central America

The countries of Central America form a long isthmus between the rest of North America and the continent of South America. It is a rugged, mountainous area with more than 250 volcanoes. On the western side, the land slopes up from a narrow coastal plain along the Pacific Ocean to the top of the mountains. On the eastern side, the land slopes more gradually from the mountaintops to a broader plain along the Caribbean Sea.

The climate of these countries varies according to altitude. From sea level up a few thousand feet, the climate is hot. At midmountain elevations, the weather is more temperate. At high levels the climate is relatively cold. The eastern coast and mountain slopes receive twice as much rainfall as the western coast and mountain slopes. Rainfall is the greatest along the eastern coast of Nicaragua.

Plant life also varies according to altitude. There are lowland rainforests along the coasts. Higher up along the mountain slopes are pine and oak forests. At the highest altitudes of Guatemala and Costa Rica are grassy areas.

The animal life is similar to that found in other parts of North America and in South America. Here are a few of the many kinds of animals found in Central America:
- boa constrictors and vine snakes
- hog-nosed skunks and raccoons
- mountain lions and peccaries
- many kinds of bats and birds

Answer these questions:

1. How many countries are in Central America? _____

2. List some ways these countries are alike. _____

_____

_____

_____

3. List some ways these countries are different. _____

_____

_____

_____

Name

Central America

Label the countries and their capital cities.

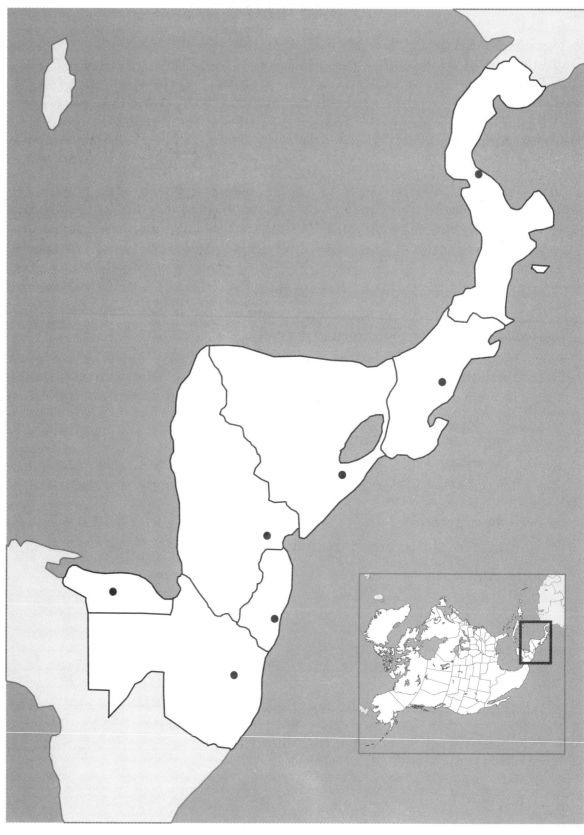

The Island Countries

The island countries of North America are sometimes called the West Indies. There are 13 independent nations in an archipelago (chain of islands) separating the Caribbean Sea and the Atlantic Ocean.

Cuba is the largest of the island countries. Many of the others are quite small. Some countries are a single island, while others consist of several islands.

Most of the West Indies islands are in the tropic zone. There are two seasons: a relatively dry season from November through May and a wet season from June through October.

1. Label these 13 independent island countries on your map of the West Indies. Check off each island's name after you label it.

☐ Bahamas

☐ Cuba

☐ Jamaica

☐ Haiti

☐ Dominican Republic

☐ Antigua and Barbuda

☐ St. Kitts and Nevis

☐ Dominica

☐ St. Lucia

☐ Barbados

☐ St. Vincent and the Grenadines

☐ Grenada

☐ Trinidad and Tobago

2. Puerto Rico is part of the United States. Find and label it on your map.

Name

West Indies

Label all islands.

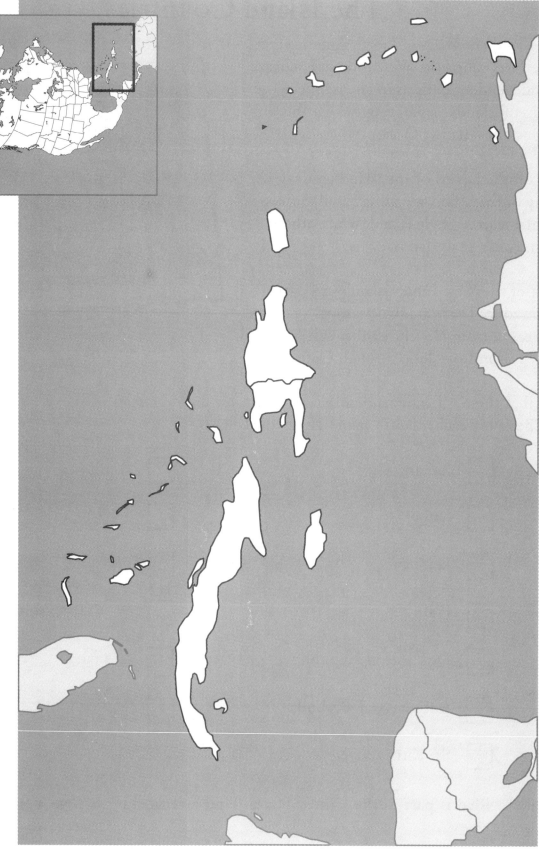

How Many People Live Here?

This graph shows the average number of people living within a square mile. Use the graph to answer the questions on page 42.

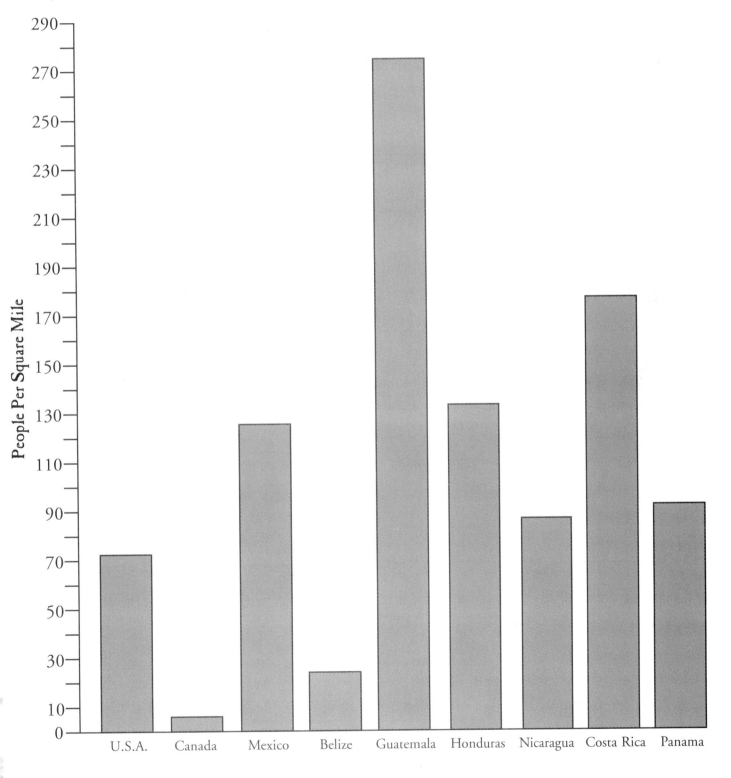

How Crowded Is It?

1. Which country on the graph is the most crowded? _____

2. Which country on the graph is the least crowded? _____

3. Find the average population per square mile for El Salvador and explain why it was not included on this graph.

   _____

4. List the countries on the chart in the order of population from the smallest to the largest.

   a. _____   f _____

   b. _____   g. _____

   c. _____   h. _____

   d. _____   i. _____

   e. _____

5. List some ways the population of a country affects the way people live.

   _____

   _____

   _____

Bonus

Find the average number of people per square mile in the island countries of North America. Use the information to make your own population graph on a sheet of graph paper.

Cuba	Jamaica	Bahamas
Dominica	Grenada	St. Kitts and Nevis
Barbados	Haiti	St. Lucia
Dominican Republic	Trinidad and Tobago	St. Vincent and the Grenadines
Antigua and Barbuda		

Name of country

Country Fact Sheet

Capital city _____

Land area _____

Population _____

Major languages spoken

_____

_____

Natural resources, crops, industries

_____

_____

Physical description of the country

_____

_____

_____

Interesting plants and animals

_____

_____

Facts about the people living there

_____

_____

_____

_____

_____

_____

_____

_____

_____

Draw the country's flag here.

North America's Resources

The activities in this section introduce students to the natural and man-made resources of North America.

Resources

Prepare for this lesson by enlarging the political map on page 23, using an overhead projector and a sheet of butcher paper. Post the map on a bulletin board.

Reproduce page 46 for each student. Assign a country to each student or small group. Explain to students that they will be looking for information about the natural resources, crops and livestock, manufactured goods, and services that create the economy of the countries of North America. (Discuss and define each of these terms before beginning the activity.) Students use atlases, almanacs, books, and the World Wide Web to locate information for the country they have been assigned and record the information gathered on their activity pages.

Create a "key" on the map using symbols agreed upon by the students. Then have students place symbols for the items on their lists in the appropriate locations on the large map of North America.

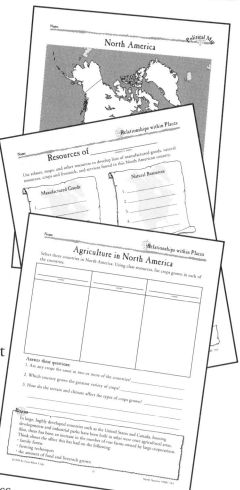

When the class map is completed, extend the lesson by asking questions such as the following:

"Which countries have the most varied economies?"
"Does (name country) have much agriculture?"
"Does (name country) have varied industry?"
"How are the products of developed countries different from the products of less-developed countries?"

Agriculture

Reproduce page 47 for each student. Have students work individually or in small groups to compare the agricultural products of three North American countries. They will use class resources to complete the activity and answer the questions. Provide time for students to share information with the class.

©1999 by Evan-Moor Corp.

North America • EMC 763

Imports–Exports

Make an overhead transparency of page 48 and reproduce page 49 for each student. Discuss the terms "import" and "export" with students. As a class, write a definition of each term. Display the transparency of Honduran imports and exports to model the activity. Have students read the chart to explain the imports and exports as listed. Show where this information is located in almanacs and/or electronic resources.

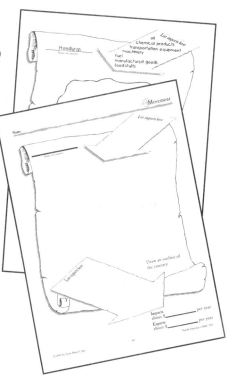

Assign a country to each student or small group. Students use class sources to find the information needed to complete their charts. Allow time for students to share the information discovered with the class.

Vacation Time

Come to North America

Tourism is a big industry in many parts of North America. Visit a travel agency to get samples of brochures and posters about trips to parts of Canada, the U.S.A., Mexico, Central America, and the Caribbean. After sharing these materials, have students develop one of the following:

- a brochure of things to do on a vacation in a part of North America
- a travel poster about one special place or site in North America
- a list of ways to be a considerate tourist
- a video advertisement encouraging people to come to a part of North America

A North American Vacation

Students are to plan a vacation in North America. They are to do the following:

- select at least one region they would like to visit
- explain why they selected the region
- research activities available in the region
- write a letter about the trip to a friend

Resources of _____
country's name

Use atlases, maps, and other resources to develop lists of manufactured goods, natural resources, crops and livestock, and services found in this North American country.

Manufactured Goods

1. _____
2. _____
3. _____
4. _____
5. _____

Natural Resources

1. _____
2. _____
3. _____
4. _____
5. _____

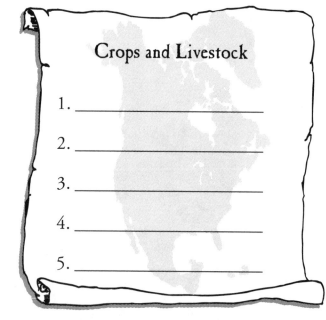

Crops and Livestock

1. _____
2. _____
3. _____
4. _____
5. _____

Services

1. _____
2. _____
3. _____
4. _____
5. _____

Bonus

Explore the types of transportation and communication used in this country. How do these two elements affect the economy of the country?

Agriculture in North America

Select three countries in North America. Using class resources, list crops grown in each of the countries.

_____ name	_____ name	_____ name

Answer these questions:

1. Are any crops the same in two or more of the countries? _____

2. Which country grows the greatest variety of crops? _____

3. How do the terrain and climate affect the types of crops grown? _____

_____

_____

Bonus

In large, highly developed countries such as the United States and Canada, housing developments and industrial parks have been built in what were once agricultural areas. Also, there has been an increase in the number of vast farms owned by large corporations. Think about the affect this has had on the following:

* family farms
* farming techniques
* the amount of food and livestock grown

Honduras
Name of country

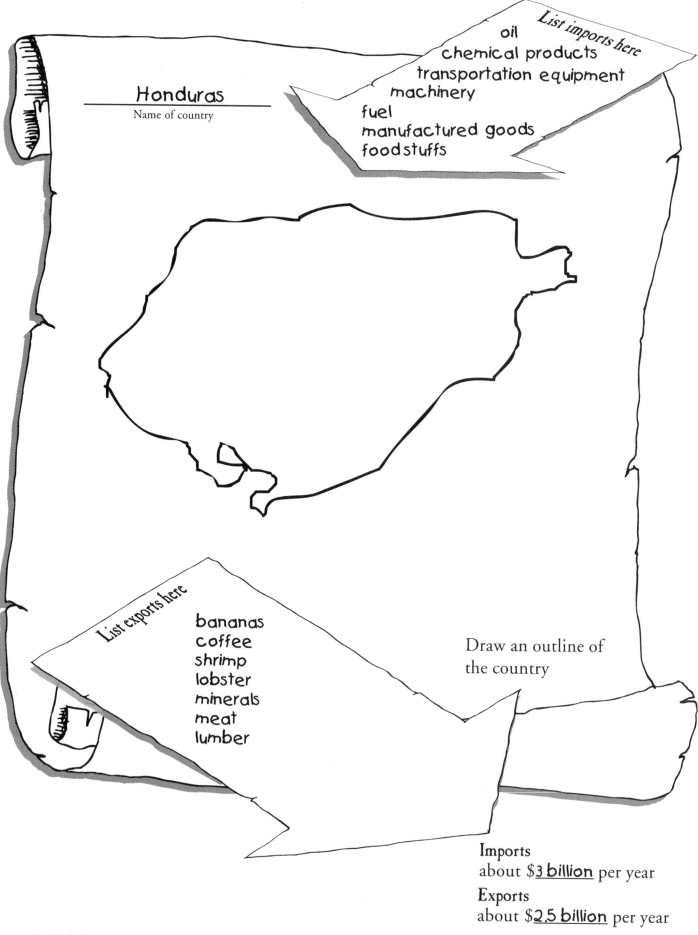

List imports here

oil
chemical products
transportation equipment
machinery
fuel
manufactured goods
foodstuffs

List exports here

bananas
coffee
shrimp
lobster
minerals
meat
lumber

Draw an outline of
the country

Imports
about $3 billion per year

Exports
about $2.5 billion per year

North America • EMC 763

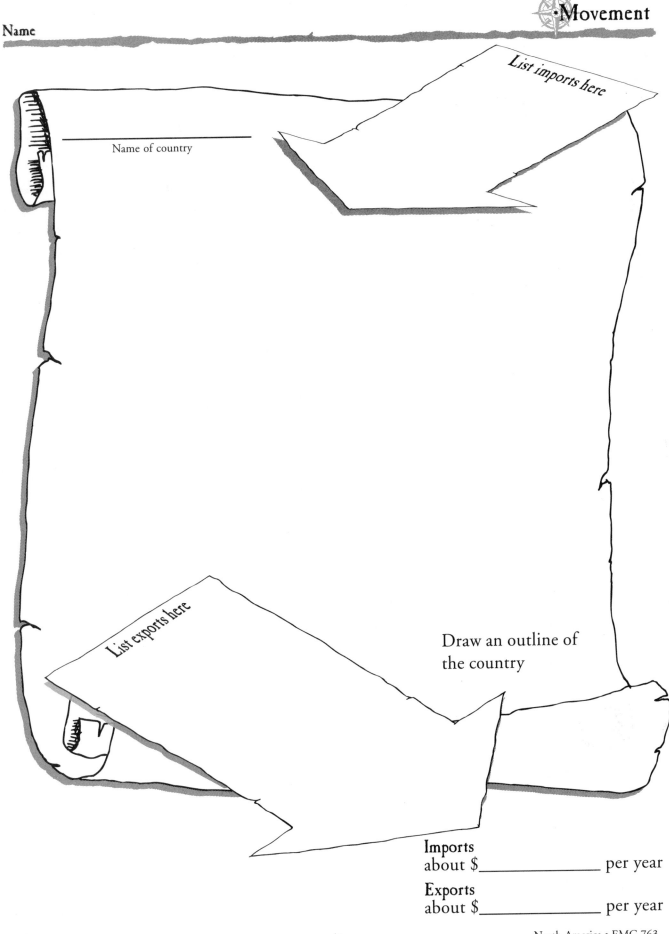

List imports here

_____
Name of country

List exports here

Draw an outline of
the country

Imports
about $_____ per year

Exports
about $_____ per year

North America • EMC 763

North American Animals

Each region of North America has its own unique animals. In this section students will learn about many of them.

Introductory Activities

Begin by challenging students to name the wild animals of North America they know. List these on a chart and write a descriptive phrase after each name.

Owl

> cougar—a large, golden-brown wild cat
> moose—a strange-looking mammal with huge antlers
> sea otter—a small sea mammal with soft fur
> musk-ox—a large mammal with long, shaggy hair

Share books or show a video about North American animals. Discuss information learned from these sources, and add new animal names and descriptive phrases to the chart.

Amazing Animals of North America

Reproduce pages 51 and 52 for each student. Have students use class resources to locate two interesting facts about each animal shown.

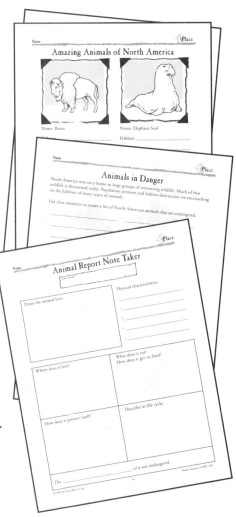

Endangered Animals

Animals in Danger

Reproduce page 53 for each student. As a class, brainstorm a list of North American animals that students think are endangered. Record these on a chart. Then have students use class resources to find animals that are actually endangered. (These are to be listed on the student's activity page.) Compare student lists to the chart. Mark out any animals on the class chart that are not endangered.

Have each student create a poster informing people about one endangered animal.

North American Animal Report

Provide each student with a copy of the note taker on page 54. Students choose one interesting North American animal and use class resources to locate the information to complete the note taker. Then students synthesize what they have learned into an oral or written report.

Amazing Animals of North America

Name: Bison

Habitat: _____

Facts: _____

Name: Elephant Seal

Habitat: _____

Facts: _____

Name: Armadillo

Habitat: _____

Facts: _____

_____

Name: Caribou

Habitat: _____

Facts: _____

_____

Name: Manatee

Habitat: _____

Facts: _____

_____

Name: Kinkajou

Habitat: _____

Facts: _____

_____

Name: Weasel

Habitat: _____

Facts: _____

_____

Name: Ptarmigan

Habitat: _____

Facts: _____

_____

Animals in Danger

North America was once home to large groups of interesting wildlife. Much of that wildlife is threatened today. Population pressure and habitat destruction are encroaching on the habitats of many types of animals.

Use class resources to create a list of North American animals that are endangered.

_____   _____

_____   _____

_____   _____

_____   _____

_____   _____

_____   _____

_____   _____

_____   _____

_____   _____

_____   _____

Bonus

What animals in your own country are endangered?

Name _____

Animal Report Note Taker

Name of animal
_____

Draw the animal here.

Physical characteristics:

* _____

* _____

* _____

* _____

* _____

Where does it live?

What does it eat?
How does it get its food?

How does it protect itself?

Describe its life cycle.

The _____ is/is not endangered.

The People

North America's People

Introduction

Invite speakers from the various countries of North America to speak to the class. Prepare students for your speakers by planning questions to ask. Appoint several students to record questions asked and answers received. Follow up the visit by writing thank-you letters.

The People of North America

Reproduce pages 57–59 for each student. Introduce the terms "ethnic groups" (having to do with a group of people who have the same race, nationality, or culture) and "cultural groups" (sharing the same customs, arts, language, etc.), but don't expect students to make the distinctions independently.

As a class, read and discuss the information. Students then use this information, plus class resources, to complete the questions on pages 57 and 58.

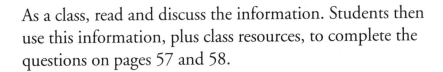

Divide the class into small groups. Assign a North American country to each group. Students explore the celebrations of their assigned country, recording the information on page 59. Allow time for each group to share what they learned. You can extend the activity by creating a calendar on a large sheet of butcher paper where students can record North American celebrations.

Life Expectancy

One of the facts students can find out about a country is the average life expectancy for a person who lives there. The figure changes every year as living conditions and health care change. Make an overhead transparency of page 60. Reproduce pages 60 and 61 for each student. As a class, read the graph and then answer the questions. (The figures are from 1997.)

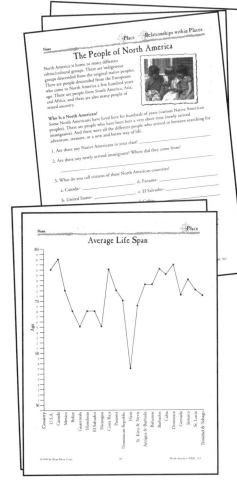

The First Americans

Reproduce page 62 for each student. Prepare for this activity by gathering books from your school and class libraries about native peoples. There are many books available about specific tribes. There are also books such as *Atlas of Indians of North America* by Gilbert Legay and *From Abenaki to Zuni: A Dictionary of Native American Tribes* by Evelyn Wolfson that give an overview of many tribes.

Invite Native Americans to speak to the class. You may have students in class with Native American heritage. These students and their families can be wonderful sources of information.

Share some of the books you collected to give students an idea of the great number of Native American groups that are a part of the population of North America. Allow students time to browse through the materials, and then ask each student to select a group to learn more about. Students record information on their note takers, and then synthesize the information into oral or written reports. Provide time for students to share what they've learned with the rest of the class.

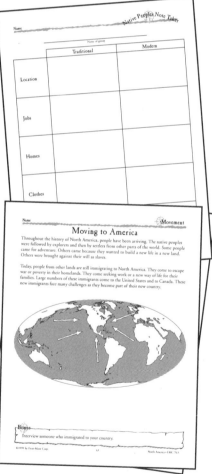

Moving to America

Immigration has been the major way in which the countries of North America have grown. Reproduce pages 63 and 64 for each student. As a class, read and discuss the information. Share a book such as *Coming to America: The Story of Immigration* by Betsy Maestro with students, and have them use the geography center to find more information about immigration. They will use this information to complete their activity pages.

The People of North America

North America is home to many different
ethnic and cultural groups. There are indigenous
groups descended from the original native peoples.
There are people descended from the Europeans
who came to North America a few hundred years
ago. There are people from South America, Asia,
and Africa, and there are also many people of
mixed ancestry.

Who Is a North American?

Some North Americans have lived here for hundreds of years (various Native American
peoples). There are people who have been here a very short time (newly arrived
immigrants). And there are all the different people who arrived in between searching
for adventure, treasure, or a new and better way of life.

1. Are there any Native Americans in your class? _____

2. Are there any newly arrived immigrants? Where did they come from?

    _____

3. What do you call citizens of these North American countries?

    a. Canada– _____   d. Panama– _____

    b. United States– _____   e. El Salvador– _____

    c. Belize– _____   f. Cuba– _____

Where Do North Americans Live?

The people of North America live in small towns, in villages, on farms, and in large
modern cities. They live along the coasts, high in the mountains, on the wide grasslands, and
in the hot desert. The culture of each country is a mixture of parts from the various peoples
that have settled there. North America is home to many different groups of people.

What Languages Do They Speak?

People in the different countries of North America speak a variety of languages. Although many people speak more than one language, each country has at least one official language. List the official languages of these countries:

1. Canada– _____

2. Mexico– _____

3. Belize– _____

4. Honduras– _____

5. Costa Rica– _____

6. Haiti– _____

7. Cuba– _____

8. Jamaica– _____

How Wealthy Are the Countries?

Canada and the United States are both highly industrial, wealthy nations. Most people in these two countries have a high standard of living. The other countries in North America vary in the degree to which they are developed. Some are increasing their amount of agriculture and manufacturing and service industries, while others are still less developed, with a poor population. Use class resources to find the average income of the people in these North American countries:

1. Jamaica– _____

2. Honduras– _____

3. Mexico– _____

4. Belize– _____

5. Haiti– _____

6. Panama– _____

7. United States of America– _____

8. Canada– _____

Celebrations

Each country in North America celebrates special events. Events such as Canadian Day are celebrated in only one country. Others such as Christmas are celebrated in all countries but in different ways. Some celebrations have spread in different countries as immigrant populations have grown. Many people in the United States attend Cinco de Mayo or Saint Patrick's Day celebrations, even though they are not Mexican or Irish.

Bonus

Think of some reasons for the differences in wealth among the various countries of North America.

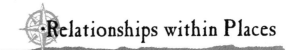
Celebrations in _____
name of country

Celebration Date	Why Celebrated	How Celebrated

Average Life Span

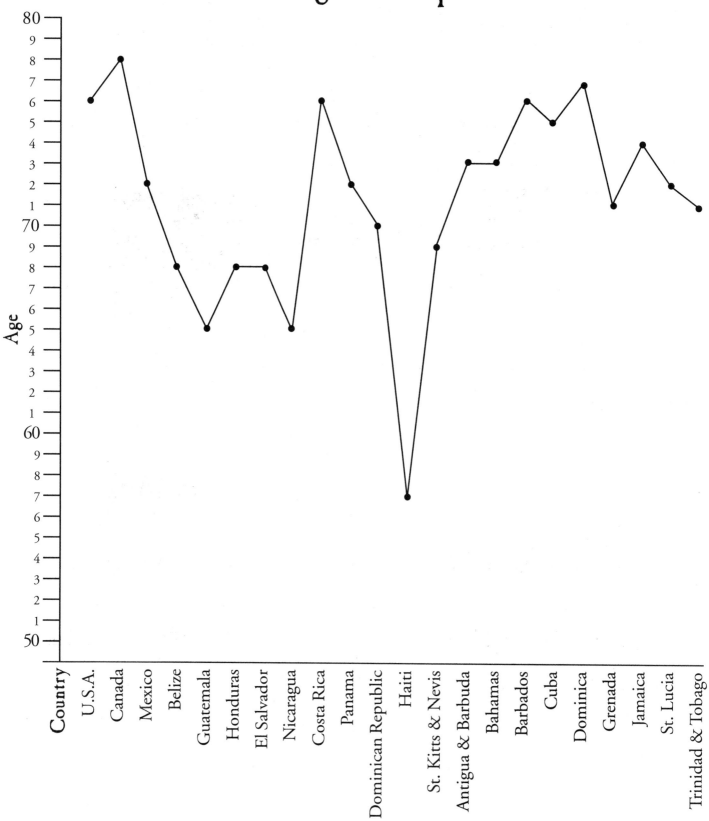

North American Life Expectancy

Use the graph on page 60 to answer these questions:

1. In which country in North America do people have the longest average life span? _____

    the shortest? _____

2. In how many countries can people expect to live longer than 70 years? _____

3. List some reasons people live longer in some countries than in others. _____

    _____

4. List the countries in the order of average life span from longest to shortest.

    1. _____   12. _____

    2. _____   13. _____

    3. _____   14. _____

    4. _____   15. _____

    5. _____   16. _____

    6. _____   17. _____

    7. _____   18. _____

    8. _____   19. _____

    9. _____   20. _____

    10. _____   21. _____

    11. _____   22. _____

5. Does the life span figure for your country mean that you will die when you reach that age?

    Explain your answer. _____

    _____

Name _____

Native Peoples Note Taker

_____
Name of group

	Traditional	Modern
Location		
Jobs		
Homes		
Clothes		
Ceremonies		

Moving to America

Throughout the history of North America, people have been arriving. The native peoples were followed by explorers and then by settlers from other parts of the world. Some people came for adventure. Others came because they wanted to build a new life in a new land. Others were brought against their will as slaves.

Today, people from other lands are still immigrating to North America. They come to escape war or poverty in their homelands. They come seeking work or a new way of life for their families. Large numbers of these immigrants come to the United States and to Canada. These new immigrants face many challenges as they become part of their new country.

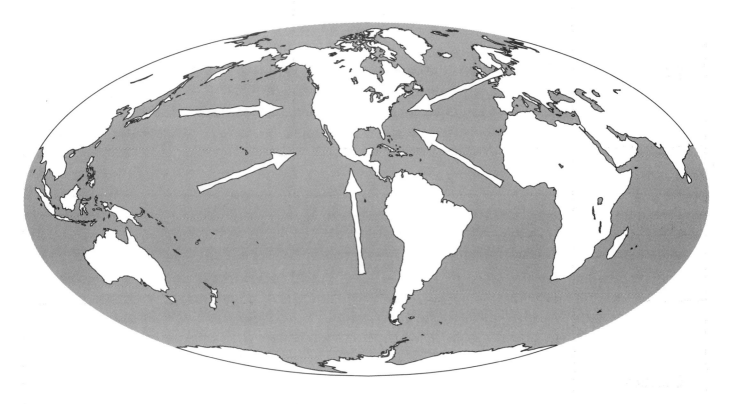

Bonus

Interview someone who immigrated to your country.

Immigrants

Use class resources to find out more about the newest immigrants to North America:

1. Where are the immigrants coming from? _____

_____

2. Why are they coming? _____

_____

3. Where in North America are they settling? _____

_____

4. What types of problems do immigrants face when they arrive in a new country?

_____

_____

_____

5. In what ways has the increased populations of non-English speaking people affected the United States and Canada?

_____

_____

_____

6. In what ways do new immigrants help a country? _____

_____

_____

Bonus

Find out about immigrants in your community. Use a map of the world to mark where they have come from.

Celebrate Learning

Choose one or all of the following activities to celebrate the culmination of your unit on North America. Use the activities to help assess student learning.

Have a Portfolio Party

Invite parents and other interested people to a "portfolio party" where students will share their completed portfolios, as well as other projects about North America.

Write a Book

A student can make a book about North America. It might be one of the following:
- an alphabet book of North American people, places, or plants and animals
- a dictionary of words pertaining to North America
- a pop-up book of the unique animals of North America

Conduct an Interview

A student can interview someone from a North American country or someone who has visited there. The interview could be in person, written about, or videotaped to share with the class.

Create a Skit

One or more students can write and present a skit about an interesting event or period in North American history.

Paint a Mural

One or more students can paint a mural showing one region of North America. A chart of facts about the region should accompany the mural.

Share an Artifact Collection

Students can bring in one or more artifacts representative of North America such as an art object. A written description of each artifact should be included in the display.

Sing and Dance

Students can present a song (perhaps the national anthem or a traditional folk song) or a dance from a region or country in North America.

Note: Reproduce this form for each student. Use it throughout the unit to record basic information. (See page 6.)

Name _____

Summary of Facts

North America

Relative location _____

Number of countries _____

Continent land area _____

Largest country by area

_____

Smallest country by area

_____

Continent population _____

Largest country by population

_____

Smallest country by population

_____

Highest point _____

Lowest point _____

Longest river _____

Largest island _____

Interesting facts about the continent's regions:

* _____

* _____

* _____

* _____

* _____

Interesting facts about the plant and animal life:

* _____

* _____

* _____

* _____

* _____

Interesting facts about the people:

* _____

* _____

* _____

* _____

* _____

* _____

* _____

* _____

Name

What's Inside This Portfolio?

Date	What It Is	Why I Put It In

Name

My Bibliography

Date	Title	Author/Publisher	Kind of Resource

What is the highest point in North America?

What is its elevation?

1

What is the lowest point in North America?

What is its elevation?

2

Name the five Great Lakes.

3

Which of the Great Lakes are in both Canada and the U.S.A.?

4

What is the largest island in North America?

What country claims the island?

5

What is the longest river in North America?

In what country is it located?

6

Which of these North American countries has the largest land area—Belize, Honduras, or El Salvador?

7

What is the name of the seaway from the Atlantic Ocean to the Great Lakes?

8

What is the name of the large range of mountains found in western Canada and the U.S.A.?

9

Search

Name the countries that make up Central America.

10

Search

What is the name of the canal connecting the Atlantic and Pacific Oceans?

11

Search

Niagara Falls is between which two lakes?

12

Search

The equator passes through which North American countries?

13

Search

By what name are the island countries in the Caribbean Sea sometimes called?

14

Search

How many countries in North America share a border with Mexico?

Name them.

15

Search

Which two states are not a part of the contiguous United States?

16

Search

Which country in North America has the largest land area?

17

Search

Which North American countries border the Pacific Ocean?

18

Name the North American countries through which the Arctic Circle passes.

19

What river forms part of the border between Mexico and the U.S.A.?

20

Name this monument in the U.S.A. and tell where it is located.

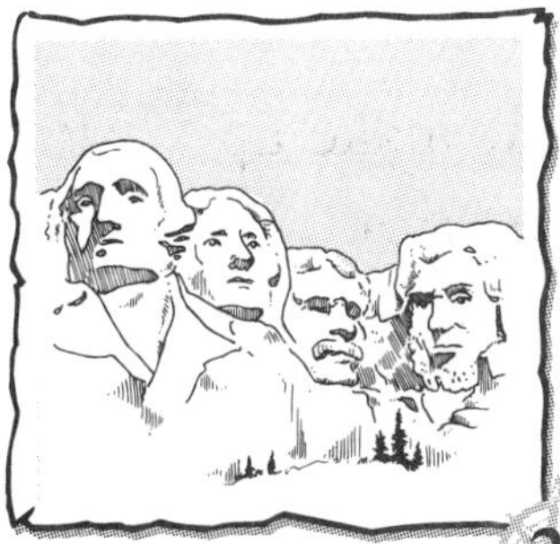

21

Name the oldest living trees found in North America.

22

Which country in North America has the smallest land area?

23

Name the largest freshwater lake in North America.

24

Name the countries these flags represent.

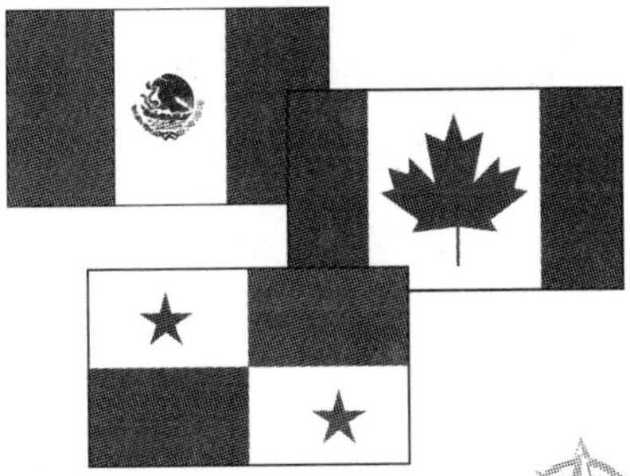

25

What is the official language spoken by the most countries in North America—Spanish, French, or English?

26

Name the sea in which the island of Cuba is located.

27

North America

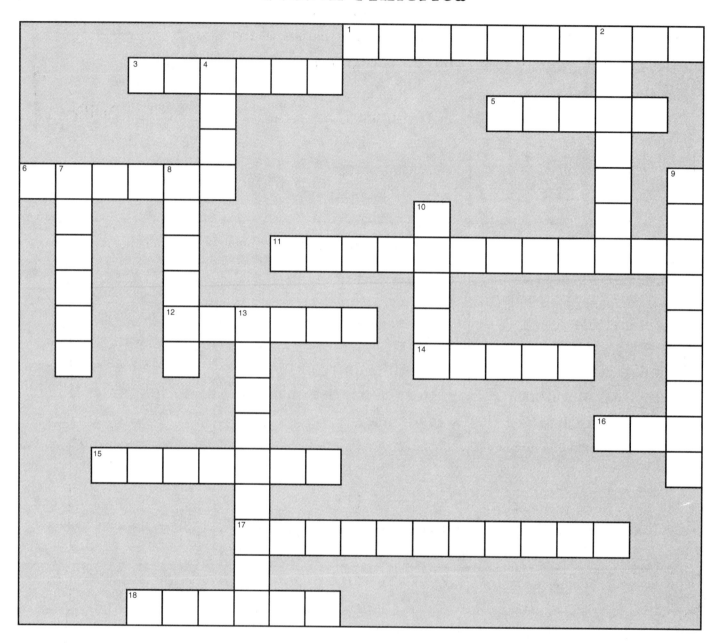

Word Box

Alaska	Grand Canyon	Niagara
Arctic	Great	Nicaragua
Canada	Hawaii	Panama
Central	Hudson Bay	Rocky
continents	Mexico	U.S.A.
Cuba	Newfoundland	Yukon

Across

1. North America is one of the Earth's seven _____
3. the North Pole is located in the _____ Ocean
5. the five lakes between Canada and the U.S.A. are called the _____ Lakes
6. the _____ Canal connects the Atlantic Ocean and the Pacific Ocean
11. the easternmost province of Canada
12. Ottawa is the capital of this large nation in North America
14. a territory in northwest Canada
15. the countries between Mexico and South America are called _____ America
16. the abbreviation for the United States of America
17. the _____ is a large canyon in the western U.S.A.
18. an island state in the Pacific Ocean that is part of the U.S.A.

Down

2. the _____ Falls are found between New York, U.S.A., and Ontario, Canada
4. a large island country located in the Caribbean Sea
7. the northernmost state in the U.S.A.
8. the country that borders the United States to the south
9. a large bay in Canada
10. the _____ Mountains extend through western Canada and the U.S.A.
13. the country between Honduras and Costa Rica

North America

```
U S E S I C O H O N D U R A S C O G B
N A A M A O I C E B E R G I I A D R A
I S M E L S A L V A D O R V S L O E H
T K O L T T X F O R E S T E L I M E A
E A O K L A D O M I N I C A A F I N M
D T S Q P R A I R I E D G G N O N L A
S C E H A I T I N U I T W D D R A A S
T H U L F C U B A Y G U L F O N C N T
A E X G U A T E M A L A R L U I A D E
T W B U F F A L O N T A R I O A N A R
E A Y U K O N I D E S E R T H L R L R
S N N O R T H Z N Z C O X S F A E B I
O L J U N G L E T S A L M O N S P E T
F A T P A N A M A X N A T I O K U R O
A K E A R T Q N I C A R A G U A B T R
M E X I C O U O N O D B U T A H L A Y
E E A R T T E T S L A E G R E T I O S
R R S K I E B A L D E A G L E P C H T
I I L K C M E P R O V I N C E A T I A
C E T T O N C A R I B B E A N K Z O T
A C O N T I N E N T J A M A I C A X E
```

Find these words:

Alaska
Alberta
Arctic
Bahamas
Bald eagle
Bay
Belize
Buffalo
California
Canada
Caribbean

Continent
Costa Rica
Cuba
Desert
Dominica
Dominican Republic
Egret
Elk
El Salvador
Forest
Greenland
Guatemala
Gulf

Haiti
Honduras
Iceberg
Inuit
Island
Jamaica
Jungle
Lake Erie
Mexico
Moose
Nicaragua
Ohio
Ontario

Panama
Prairie
Province
Quebec
Salmon
Saskatchewan
State
Territory
Texas
United States of America
Utah
Yukon

Bonus

Circle the names of the North American countries on the list.

Glossary

absolute location (exact location)–the location of a point that can be expressed exactly, for example, the intersection of a line of longitude and latitude.

altitude–the height of a thing above a given reference point; the height of a thing above sea level.

Amerindian–a name given collectively to the indigenous peoples of the Americas.

archipelago–a large group or chain of islands.

Arctic–the area at or near the North Pole.

Arctic Circle–an imaginary line circling the globe at 66.5°N latitude.

bay–part of a sea or river extending into the land.

cape–a piece of land that extends into a river, a lake, or an ocean.

capital–a city where a state or country's government is located.

cardinal directions–the four points of a compass indicating north, south, east, and west.

climate–the type of weather a region has over a long period of time.

compass rose–the drawing on a map that shows the cardinal directions.

continent–one of the main landmasses on Earth (usually counted as seven—Antarctica, Australia, Africa, North America, South America, Asia, and Europe).

culture–the shared way of life of a people including traditions, beliefs, and language.

equator–an imaginary line that circles the Earth midway between the north and south poles, dividing it into two equal parts.

ethnic group–a group of people sharing the same origin and lifestyle.

gulf–a portion of an ocean or sea partly enclosed by land.

hemisphere–half of a sphere; one of the halves into which the Earth is divided—western hemisphere, eastern hemisphere, southern hemisphere, or northern hemisphere.

indigenous–native to an area; originating in the region or country where it is found.

immigrant–a person who has come from one country to live in a new country.

isthmus–a narrow strip of land with water on both sides, connecting two larger bodies of land.

landform–the shape, form, or nature of a physical feature on Earth's surface (mountain, mesa, plateau, hill, etc.).

latitude–the position of a point on Earth's surface measured in degrees, north or south from the equator.

longitude–the distance east or west of Greenwich meridian (0° longitude) measured in degrees.

manufacture–to make a useful product from raw materials.

meridian–an imaginary circle running north/south, passing through the poles and any point on the Earth's surface.

North Pole–the northernmost point on Earth; the northern end of the Earth's axis.

plain–a flat or level area of land not significantly higher than surrounding areas and with small differences in elevation.

plateau–an area of land with a relatively level surface considerably raised above adjoining land on at least one side.

population–the total number of people living in a place.

prime meridian (Greenwich meridian)–the longitude line at 0° longitude from which other lines of longitude are measured.

province–one of the main administrative divisions of a country.

rainforest–a dense forest found in wet, tropical regions.

relative location–the location of a point on the Earth's surface in relation to other points.

reservation–land set aside for a special purpose such as an Indian reservation.

reserve–public lands set aside for a special purpose such as an animal reserve.

resource–substances or materials that people value and use; a means of meeting a need for food, shelter, warmth, transportation, etc.

rural–relating to the countryside.

scale–an indication of the ratio between a given distance on the map to the corresponding distance on the Earth's surface.

service–activities that are paid for although nothing is produced, for example, banking or tourism.

South Pole–the southernmost point on Earth; the southern end of the Earth's axis.

strait–a narrow passage of water connecting two large bodies of water.

symbol–something that represents a real thing.

territory–a region of a country not admitted as a state or province but having its own legislature and governor.

Tropic of Cancer–an imaginary line around the Earth north of the equator at the 23.5°N parallel of latitude.

tundra–vast, treeless plains of the Arctic.

urban–relating to cities.

Answer Key

1. a wide plateau in central North America
2. from northwest Canada to southern U.S.A.; from the Rocky Mountains to the Canadian Shield
3. bison, antelope
4. Crops cover the land where wild grasses once grew, and domestic animals are found where wild herds used to roam.
5. Answers will vary, but could include wheat and corn.
6. Answers will vary, but could include cattle and sheep.

page 18
1. reaches from central New Mexico to northwestern British Colombia
2. the Great Plains
3. Answers will vary, but could include: Grand Teton, Yellowstone, Yoho, and Glacier.
4. Mount Elbert–14,433 feet (4399 m) elevation
5. Answers will vary, but could include caribou and moose.
6. Answers will vary, but could include pumas, coyotes, and opossums.

page 19
1. arctic plains of the north; permafrost under surface soil
2. harsh winters, low temperatures, little rain, short summers
3. The permafrost keeps the melted snow and ice from soaking down into the ground, so pools of water form on the surface.
4. lichen and mosses
5. Answers will vary, but could include: musk-ox, caribou, reindeer, wolves, arctic foxes, and polar bears.

page 25
U.S.A.–Washington, D.C.
Canada–Ottawa
Mexico–Mexico City
Guatemala–Guatemala City
El Salvador–San Salvador
Nicaragua–Managua
Costa Rica–San José
Belize–Belmopan
Honduras–Tegucigalpa
Panama–Panama City

page 27
1. Ottawa
2. 10 provinces
3. 3 territories
4. English and French
5. Toronto, Ontario
6. constitutional monarchy with a democratic parliament
7. Roman Catholic
8. Answers will vary.

page 30
1. Whitehorse
2. Victoria
3. Toronto
4. Regina
5. Halifax
6. Ottawa
7. Edmonton
8. Quebec
9. Saint John's
10. Calgary

1. Montréal–46°N, 73°W
2. Vancouver–49°N, 123°W

page 31
1. Washington, D.C.; District of Columbus
2. 50
3. Hawaii and Alaska
4. English
5. largest–California; smallest–Wyoming
6. 50 stars stand for the 50 states; 13 stripes for the original 13 colonies
7. Answers will vary.

page 35
1. Mexico City; Federal District
2. 31 states
3. Mexico City
4. Spanish
5. Roman Catholic
6. 1429 miles (2300 km)
7. Federal republic with centralized government
8. Answers will vary.

page 37
1. 7 countries
2. Answers will vary, but could include: all small countries, all south of the Tropic of Cancer, all grow crops, and all touch an ocean or sea.
3. Answers will vary, but could include: speak different languages, different plants and animals, and different histories.

page 42
1. Guatemala
2. Canada
3. 697 per square mile; There are too many people to fit on the graph the way it is made.
4. a. Canada   d. Nicaragua   g. Honduras
   b. Belize   e. Panama   h. Costa Rica
   c. U.S.A.   f. Mexico   i. Guatemala
5. Answers will vary, but could include: how crowded living areas are, the amount of space needed for food production, the amount of water and air pollution.

page 46
Answers will vary, but could include:
Manufactured Goods–vehicles, machinery, chemicals, steel
Natural Resources–petroleum, fish, timber, minerals
Crops & Livestock–wheat, bananas, cotton, beef, poultry
Services–banks, transportation, schools, medical care

page 57
1. Answers will vary.
2. Answers will vary.
3. a. Canadian   c. Belizean   e. Salvadoran
   b. American   d. Panamanian   f. Cuban

page 58
1. English and French
2. Spanish
3. English
4. Spanish
5. Spanish
6. French and Creole
7. Spanish
8. English

page 58 (continued)
(Answers will depend on the source used. These figures are from a 1998 World Almanac.)
1. $3,200
2. $1,980
3. $7,700
4. $2,750
5. $1,000
6. $5,100
7. $27,607
8. $24,400

page 61
1. Canada/Haiti
2. 16 countries
3. Answers will vary, but could include: better nutrition, less pollution, and greater access to medical care.
4. 1. Canada
   2. Dominica
   3. U.S.A.
   4. Costa Rica
   5. Barbados
   6. Cuba
   7. Jamaica
   8. Antigua & Barbuda
   9. Bahamas
   10. Mexico
   11. Panama
   12. St. Lucia
   13. Grenada
   14. Trinidad & Tobago
   15. Dominican Republic
   16. St. Kitts & Nevis
   17. Belize
   18. Honduras
   19. El Salvador
   20. Guatemala
   21. Nicaragua
   22. Haiti
5. No, this is an average figure. Some people will die sooner and some will live much longer.

page 69
1. Mt. McKinley in Alaska, U.S.A.–20,320 feet (6194 m) above sea level
2. Death Valley in California, U.S.A.–282 feet (86 m) below sea level
3. Superior, Michigan, Huron, Erie, Ontario
4. Superior, Huron, Erie, Ontario
5. Greenland, Denmark
6. Mississippi River in the U.S.A.
7. Honduras
8. St. Lawrence
9. Rocky Mountains

page 70
10. Belize, Guatemala, El Salvador, Honduras, Nicaragua, Costa Rica, Panama
11. Panama Canal
12. Lake Erie and Lake Ontario
13. none
14. West Indies
15. 3–U.S.A., Guatemala, Belize
16. Hawaiian Islands and Alaska
17. Canada–3,849,674 square miles (9,970,609 sq km)
18. Canada, U.S.A., Mexico, Guatemala, El Salvador, Honduras, Nicaragua, Costa Rica, Panama

page 71
19. U.S.A. (in Alaska) and Canada
20. Rio Grande
21. Mt. Rushmore in South Dakota
22. bristlecone pines
23. St. Kitt and Nevis–104 square miles (269 sq km)
24. Lake Superior
25. Mexico, Canada, Panama
26. English
27. Caribbean Sea

page 72

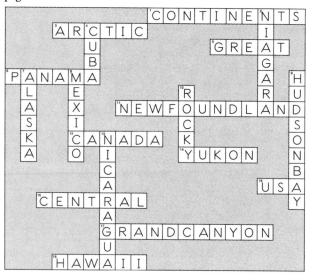

page 74

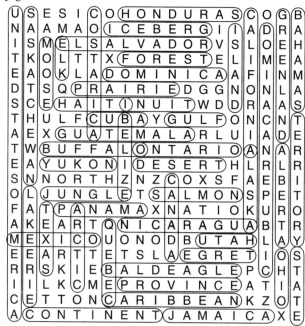

Bibliography

Books about North America

America: Land of Beauty and Splendor; Reader's Digest, 1992.

Atlas of Indians of North America by Gilbert Legay; Barrons Educational Series, Inc., 1995.

Belize (Enchantment of the World Series) by Marion Morrison; Children's Press, 1996.

Canada (Major World Nations) by Kevin Law; Chelsea House Publishers, 1999.

Canada: The Culture by Bobbie Kalman; Crabtree Publications, 1993.
(Also available—*Canada: The Land* and *Canada: The People*)

Coming to America: The Story of Immigration by Betsy Maestro; Scholastic Inc., 1996.

From Abenaki to Zuni: A Dictionary of Native American Tribes by Evelyn Wolfson; Walker Publishing Company, Inc., 1988.

Let's Explore the Northeast (America, This Land is Your Land) by Jill C. Wheeler; Abdo & Daughters, 1994. (Also in this series: *The West, The Midwest and the Heartland, The Pacific West, The Southeast and the Gulf States*)

Mexico (Country Fact Files) by Edward A. Parker; Raintree Steck-Vaughn Publishers, 1996.

Mexico (Major World Nations Series) by Jack Rummel; Chelsea House Publishers, 1998.
(Many other North American countries are part of this series.)

Mexico and Central America (Places and People) by Marion Morrison; Franklin Watts, 1997.

The Big Book of America by Roger Hicks; Running Press Book Publishers, 1994.

West Indies (World in View Series) by Ron Ramdin; Steck-Vaughn Library Division, 1991.

General Reference Books

(Maps and atlases published before 1997 may not have the latest changes in country names and borders, but they will still contain much valuable material.)

Atlas of Continents; Rand McNally & Company, 1996.

National Geographic Concise Atlas of the World; National Geographic Society, 1997.

National Geographic Picture Atlas of Our World; National Geographic Society, 1994.

The New Puffin Children's World Atlas by Jacqueline Tivers and Michael Day; Puffin Books, 1995.

The Reader's Digest Children's Atlas of the World; Consulting Editor: Colin Sale; Joshua Morris Publishing, Inc., 1998.

The World Almanac and Book of Facts 1998; Editorial Director: Robert Famighetti; K-III Reference Corporation, 1997.

Technology

CD-ROM and Disks

Encarta® Encyclopedia; ©Microsoft Corporation (CD-ROM).

MacGlobe & PC Globe; Broderbund (disk).

Where in the World Is Carmen Sandiego?; Broderbund (CD-ROM and disk).

World Fact Book; Bureau of Electronic Publishing Inc. (CD-ROM).

Zip Zap Map; National Geographic (laser disc and disk).

Websites

For sites on the World Wide Web that supplement the material in this resource book, go to http://www.evan-moor.com and look for the Product Updates link on the main page.

Check this site for information on specific countries:

CIA Fact Book–www.odci.gov/cia/publications/factbook/country-frame.html